Email Marketing

A Step-by-Step System to Build Passive Income Using Email Marketing

DISCLAIMER

Copyright © 2018

All Rights Reserved

No part of this book can be transmitted or reproduced in any form including print, electronic, photocopying, scanning, mechanical or recording without prior written permission from the author.

While the author has taken the utmost effort to ensure the accuracy of the written content, all readers are advised to follow information mentioned herein at their own risk. The author cannot be held responsible for any personal or commercial damage caused by information. All readers are encouraged to seek professional advice when needed.

Table of Contents

What is Email Marketing? *9*
    Different kinds of email marketing 11
The Benefits of Email Marketing *12*
    A targeted list of customers 12
    Automated marketing process 13
    Cost-effective 14
    Personalized engagement 15
    Simple to create 16
    Easily measurable 16
    Highly scalable 18
    Huge profits 20
How to build an Email List? *23*
    FIve steps of creating an email list 23
        Step-1 Finding your market 23
        Step-2 Knowing your market 24
        Step-3 Generating your automated response sequence 24
        Step-4 Placing an email signup landing page 25
        Step-5 Promote your email signup page 26
    Tips when generating your email list 26
Segmenting Email Lists *32*
    Best practices for segmenting email lists 32

Email list segmentation tool 33

New email subscribers 33

Interest segmentation 34

Service preference segmentation 34

Rate of engagement 35

Lead segmentation 36

Opened, but not converted 36

Purchase history 37

Buyer's cycle and frequency 37

Desktop and mobile customers 38

Subscribers who help in promotion 39

The profession of your subscribers 39

Gender segmentation 40

Segmenting for education level 40

Segmenting according to income 41

Qualifying your Subscribers *42*

How to qualify your subscribers? 42

Site visitors who permitted you to send emails 43

Subscribers who filled a form on your landing page 43

Business cards you obtain from office visitors and trade events 44

Members of your social media community or forum 44

Customers who have bought a product or a service from you 45

Emails available in your address book 45

The list you have rented or purchased 46

Clean your email list 46

Building a Relationship with Your List *48*

How can you achieve all three factors?! 49

Become a consistent and disciplined email marketer 49

Share a variety of content that matches subscribers' interests 50

Make them click every email with a dazzling headline 51

Optimize emails for user convenience 52

Don't just try to sell 52

Combine multiple affiliate deals in emails 53

Write interactive call-to-actions 54

Use impressive rewards 55

Always publish on-time 55

Avoid sending too many emails every day 56

Always send easily consumable information-packed emails 57

Choose your autoresponders wisely 58

Monitor results of your efforts and modify 58

Make Money with Your List *60*

Indulge in affiliate marketing 60

Offer advertising space in your emails 61

Sell your products 61

Rent email lists to others 63

Follow a proven promotion path 63

Match sales approach with list segmentation 64

Use an up-selling approach 65

Use down-selling approach 65
Re-target "opened, but not converted" segment 66
Sell premium content through a paid subscription 67
Start one-on-one consulting or coaching 67
Rules that will help you make money consistently through your email list 69
    Rule#1 Pick your niche and stick to it 69
    Rule#2 Make your pitch valuable 70
    Rule#3 Define a path towards sales 71

How to convert every subscriber into a customer? 72
    1. A well-defined sequence 72
    2. Make buying easiest through your service 73

Conclusion *74*

References *77*

What is Email Marketing?

In one line, email marketing is how you send advertising and promotional messages through emails.

Emails provide direct access to a vast market. Almost every person uses an email account. Hence, you can use this platform for cheap, fast and effective marketing. It allows you to connect businesses, products, and services to relevant consumers. And earn in the process too!

Email is known among marketers as a highly flexible tool. You can create simple or flashy messages and send it to an extensive network with a single click. Emails can contain multimedia, texts, videos, links, images and many other features. The message generation depends on marketing goals. Some messages require simple texts, while others need links, images, and other flashy elements.

Targeted marketing is the most significant benefit of email marketing. Other tools such as television, radio, or even print don't provide a precise demographic to target. However, this is not same as in email marketing. Personalized messages are created to focus different sets of consumers or individuals. Targeted lists are generated that includes past customers who present higher chances to respond to specific messages. Hence, email marketing gives high ROI to the campaign managers.

Variety of tools, techniques, and efforts are required to enhance the pace of email marketing further.

Different kinds of email marketing

- **Direct emails**

As the name suggests, these types of emails provide direct information regarding a product, promotional offers or sales. Customers receive these messages along with a call-to-action and links. An interested customer then uses the connection to obtain products, coupons or other services.

- **Newsletters**

Many customers subscribe to news, information and other messages. Companies send regular emails to such customers in the form of newsletters. Email newsletters enhance relationship building process to improve brand loyalty and customer retention. Marketers create messages in a conversational format to engage audiences towards the provided news.

- **Transactional Emails**

These messages indulge customers in some kind of action. A customer can provide purchase confirmation or trigger some other transactional action. Marketers generally blend a sales pitch with transactional emails to engage customers in other products or services.

The Benefits of Email Marketing

A correct approach to email marketing can help you reach thousands of consumers on a weekly basis. It is all about creating a system in the beginning. Eventually, you become capable of sending thousands of messages consistently. Thus, it provides with multiple advantages such as:

A targeted list of customers

If you follow a systematic approach, email marketing always provides you relevant traffic. Relevant traffic means that you send messages to customers who are most likely to buy a product or services you are promoting.

The process of email marketing starts with list creation. You generate multiple lists based on demographics that belong to product niche you have in mind. Your content, sales pitches, product promotions and other messages attain a directed pathway.

Hence, it would not be wrong to say that you never waste your time when it comes to email marketing. Relevancy lies at the core of this marketing method.

Automated marketing process

With advanced tools present, different steps of email marketing have become entirely automated. Email creation, bulk delivery, and follow-up, everything is possible with a few clicks these days. Hence, you don't have to spend too much time managing everything.

Many tools help you create customized messages. You can update message creation processes and design a sequence around your email marketing strategy. For instance, you can define daily, weekly or monthly emails for the list you create.

Different automation processes are leveraged to create brand awareness. Then, you can move forward and design engaging messages for more interested customers.

Cost-effective

The return on investment is so high that email marketing attracts all kinds of businesses and individual marketers. You already know how automation empowers your ability to send emails without wasting any time. The same system also allows you to save a lot of money in the process.

You don't have to attain a marketing certification to start earning through email marketing. It eventually concludes with a few simple steps and the organized approach you generate. Then, you can receive a great deal of money annually.

It generally requires less than $10 per month to start your marketing process. You can easily register with email sites and use collected email IDs to send messages to consumers.

If you think it through, less than 10 dollars will allow you to create an automated money earning email marketing process. And that cycle will exist forever and let you gain hundreds of dollars in return. There is no better method of creating this passive sourece of money, with such a small investment.

Personalized engagement

In any form of marketing, personalized messages are considered most effective. However, marketers usually worry about the hard work required sending personalized messages to thousands of consumers.

Personalized engagement is not difficult nowadays. Most servers provide a customizable system that allows you to include receiver's name automatically.

For example, your message can start with:

Hey, (receiver's name), hope you're doing well.

Adding the name creates a personalized engagement, which enhances the chances of a response. The receiver feels connected to the message and opens an email to go through the content written inside.

With the right tools, you have to spend a few seconds to design a personalized email sequence and send messages to your consumers.

Simple to create

Message creation is effortless in email marketing. You don't need any technical knowledge or a big team by your side. Appealing templates, images, videos and other media resources are readily available in the market. You can collect these media on your own to create beautiful content. However, many marketers use text emails successfully with minimal media to engage people.

In emails, content is the most crucial part. All your receivers desire information that is valuable and useful. Hence, you can utilize everyday text to generate valuable messages. You don't even have to be technically savvy. There are tools in the market that allow you to drag and drop to design your emails in simple steps.

A few minutes in front of your laptop is enough to generate a sales funnel that works on its own. In return, you create a source of income for your whole life.

Easily measurable

For a marketer, it is a necessity to stay on the right track. Your whole business cycle depends on how directed your marketing strategy is. Email marketing processes are highly beneficial for tracking. Advanced software present valuable data associated with a campaign's success rate.

You can easily find out about impressions, CTR or click through rate and conversions as well. These are the most important data structures that help you monitor your email marketing approach. Changes in data allow you to take immediate actions and improve your strategies for better results.

The definition and workability of analytics is fathomable even for a person who is not a data scientist. These are simple numbers that come with a well-defined format. Most tools give a visualized form of data, so that, anyone can understand it. Eventually, you become advanced and start reading analytics faster. You can align your subscribers' growth with click-through rates obtained daily.

Similarly, you can choose tools that help you compare conversions based on location, time, day and other factors. Having all data visible at one place saves hours of effort.

Highly scalable

No other marketing method connects you to a global market. Sending emails to thousands of audiences all over the world. Social media is also there with similar reach. However, there is no guarantee whether your audience is reading your message or not.

Email Marketing offers the much-needed scalability to enhance your earnings with time. There is entirely no limit to how much you can grow with email marketing. This massive scalability attracts people from all backgrounds.

Your list keeps on growing, and that also improves your ability to make money on a grand scale. It is true that you need show patience and determination to attain more than 1 million subscribers. But that is not too difficult if you design a proven approach and stay consistent. A single individual can generate a massive list of emails within 12 months.

After generating your list, you have to sit back and let the system work for you. One day of work completes your task and creates a cycle of immediate income and high returns.

There are several tricks to keep improving the scalability of your email marketing:

- Regularly verify email addresses available on your list. Remove and replace the ones that are not working. Accuracy is the key to consistent scalability.
- Separate your messages from other promotional emails. Design a personalized message for every recipient. It will increase CTR and also positively impact conversion rates.
- Emojis are highly popular these days. You can incorporate them to make your message look more fun and entertaining. The subject line is a great spot to place emojis to stand out quickly.
- Experiment with different delivery times and compare impressions and conversions. This way, you can modify your system for maximum open rates.
- You can even try creating a separate list of people who are more responsive to your messages. It is a great approach to design your marketing messages around those confirm sales.
- Customize templated for a unique appearance. Make sure that those templates are responsive, as most people open their emails on their smartphones.
- Your email needs to have simple-to-understand CTA. Your subscribers shouldn't feel confused about

what you desire from them. Short emails followed by a clear CTA are perfect to get more responses.
- Give yourself time and let your list grow. Initial stages ask for a little bit of work. But then, you can see regular growth in responses from recipients.

Huge profits

An email marketer receives vast earnings with minimal efforts. If you carefully look at all the benefits as mentioned above, they all point towards passive income. Email marketing enables you to create lucrative income on a regular basis.

Think about the investments required for implementation. As mentioned earlier, you have to spend a minimum amount every month. The implementation process is extraordinarily cheap and takes no effort. So, you feel safe regarding investment in this passive income business.

Now, the gradual growth of your email list keeps offering a steady rise in your income. Eventually, you reach a large size of the list that takes your income to an unimaginable scale. You can expect profits up to $1000 every month if new subscribers keep on coming.

With an automated process, it is possible to add more than 500 subscribers every week. These subscribers increase the number of sales you make. As a result, your overall monthly income improves. It is all happens with $10 to $50 investment every month.

There are a variety of ways to leverage your subscribers for profits:

- Sell your product or service such as training guides, eBooks, online courses, virtual coaching or other services.
- Generate an affiliate connection to sell products or services on behalf of other businesses.
- Provide exclusive products to people who are interested in that product category.
- Modify your offers based on the interests of customers you have on your email list.
- Monetize streamlined messages to get multiple sales from customers. Seasonal and perishable products increase your chances of making repeated sales.
- Send follow-up messages requesting word-of-mouth from your subscribers.

If you can create an appealing email copy, all these methods are excellent to achieve more profits through email marketing. Most importantly, study your subscribers and their buying behavior. You can use these insights to write more effective email copies and generate more profits.

How to build an Email List?

Till now, it has become visible that the impact of your email marketing system depends on the email list. A great list of email IDs includes correct information regarding target market.

Five steps of creating an email list

Here are five most valuable steps for building an email list for passive income.

Step-1 Finding your market

To make money, you have to find a market first. Most people create their business through blogging platforms. The niche of your blog can give you a direction and help you generate your business prospect where you lead as an expert.

Just like any other market, an online market also demands specific traits from a marketer. It is your responsibility to understand those needs. You can't just create a list of random people with their email IDs. It is necessary that they are a consumer of your products. Hence, you should narrow down your market based on your expertise, your products, and services.

Step-2 Knowing your market

You need an in-depth knowledge of your market to reach out to them in an effective manner genuinely. Your list of subscribers continuously grows when you understand and act according to what your market wants.

When understanding your market, focus majorly on their buying behavior. For instance, some niches attract more customers with low-priced products. However, in other niches, you can easily convince people to pay high prices for a product. It all comes down to how that market works and the purpose of subscribers when they opt for your list.

Step-3 Generating your automated response sequence

The conclusions attained through market analysis will help you define your product. This product can be an incentive, discount, content or any other item to attract people. Now, the next thing you need to worry about is creating an automated sequence for the response.

An automated response sequence includes a series of emails that you design for subscribers. These emails should contain valuable messages and free information followed by promotions.

Email list building is not just about gaining email IDs. You have to present your emails as a valuable source of services in the initial stages. Then, you can pitch promotions and get conversions.

Remember, that you don't want free service seekers in your list. Hence, blend your free content or services with promotions. This will prepare your subscribers to become a customer.

Step-4 Placing an email signup landing page

You need a robust page that links your market to your email facility. This page is called a landing page that you place on your blog or site. A landing page provides a bright idea of what you need from your visitors. However, a successful implementation requires following qualities:

- Create a headline that appeals to the visitor.
- Choose benefit-oriented adjectives in your headline.
- Describe subscription benefits in detail via short bullet points.
- Crisply present your product with a short description.
- End with a simple call-to-action.

Step-5 Promote your email signup page

You have online as well as offline methods of promoting your email signup page. You already understand your target market. So, use that knowledge to market your email signup page through social media, blogging, forum posting, direct mailing, community events and other methods.

Tips when generating your email list

Here are all the right ways to build your email list:

- **Connect through your blog**

A blog is a great choice to create a huge community online. In fact, you can directly connect with potential customers and audiences on a regular basis. Plus, it offers you an opportunity to build a substantial email list for your target market.

Consistent blogging with a clear call-to-action will make your readers more interested in signing up for email messages. An active blog can help you create a massive list of emails in a short period.

- **Gather from social media**

Social media platforms provide a huge market where you can connect and build relationships. These interactions allow you to find new audiences who might be interested in your product or service.

You can offer solutions to their problems and encourage them to sign up for your answers. These emails will be a great addition to your targeted list.

- **Leverage telemarketing**

If your business or job allows you to interact with many people, it is an excellent source of emails for you. Whenever you call a customer, ask if he or she would prefer getting personalized messages via email. You can provide information regarding your goals with the email list. However, focus on customer-oriented goals during this discussion. Tell them, how joining your email list would help them on a regular basis. This will enhance the chances of getting more and more subscribers.

- **Ask for business cards**

You regularly meet professionals who seem like they would prefer your services. Make sure that you get their business cards and exchange it with yours. These cards will help you create a list of emails that belong to highly targeted prospects.

If you have an office, ask your visitors to leave their business cards. You can give some incentive to boost card collections.

- **Get a software**

Many email list building tools are available out there. Good software allows you to find email data targeted to your service or product directly. You can start by researching multiple tools and choose one that matches your needs.

- **Offer coupons or discounts**

People are more likely to join your email list if they get valuable offers in return. And nothing seems more valuable than a discount coupon. Almost every online business model leverages this technique. You can provide discounts in return for joining your email list. This will add a considerable number of people on your list.

- **Promise to deliver free content**

Free content is another reason why many people like to join. This is an affordable alternative to discount coupons. However, it is essential that you provide a small piece of content only. But this content needs to be valuable and directly associated with your products and customers. You can create articles, ebooks, videos and other types of content pieces for your email list.

- **Create an engaging form on your site**

A site or blog can have a separate page or popup page including sign up forms. You can place this form as a link right in the navigation bar on your site. Or, add this to your site menu.

- **Give email specials to your online community**

With blogs and sites, you can offer content that engages a vast population. But that keeps your community limited to your platform.

And as you start getting people to sign up for your email list, you start offering them email specials. These specials are services, content, products or discounts that will be available to your email subscribers only. With this, you can encourage your subscribers to share your services to others as well.

- **Reply to blog or site comments**

Two-way communication through your site or blog is way more useful for gaining email subscribers. You should answer questions on blogs, forums, and websites. Do it on your blog or site along with other sites that allow commenting. Q&A forums are highly active for interactions. You provide a solution to a problem and get a chance to pitch your email service to prospects.

- **Let word-of-mouth work for you**

A continuous cycle of email subscribers becomes possible when you leverage your current subscribers. Every time you send an email, include a request for sharing services. Your subscribers will share your services with their family and friends. Hence, you will get more emails on your list.

Remember, that your mail should contain a sign-up form. This way, every recipient of that mail gets a chance to opt into your emails.

- **Charm with multiple subscription choices**

More choices make people more interested. And you can use this to your email list advantage. You can create a variety of content associated with your product or service. Then, design multiple subscription options for all different content categories. This way, you can charm site visitors by offering them multiple choices. They can choose to subscribe for one of your content categories and gain regular emails that are relevant to their needs.

Along with the choices, you can also give freedom to your subscribers concerning email frequency. Ask them whether they like to have emails on a weekly basis, monthly basis or every time you have something new to offer. It is a great approach to give a sense of control to your audiences.

- **Give away exclusive courses on your expertise**

Fitness trainers, teachers, financial advisors, all experts have a great deal of knowledge gained through their experience. If you have such expertise, leverage it to enhance your email list.

Create a free course around your expertise. You can offer a weight loss course, cooking class, or any other. Then, start promoting this course through your blog, site, social media, forums and other online platforms. Make sure that you provide this free course only via email subscription. Hence, a person will get your particular course only when he or she subscribes to your list.

Segmenting Email Lists

Segmentation of email lists is a well-known technique used by email marketers to define highly target groups. The process includes dividing an email list into multiple segments. As an email marketer, you should find out specific needs of your subscribers. This understanding helps to choose right groups for right email campaigns. Hence, you get more conversions through your email distributions.

Best practices for segmenting email lists

Every email marketer sells a unique product, service or idea which is why everyone needs a particular segment of the market to cater. However, the pathway to segmentation goes through similar points. Here are major steps that you should incorporate to segment your email lists successfully:

Email list segmentation tool

Thousands of emails on your list can seem difficult to segment manually. So, finding an email list segmentation tool seems a logical move. You can choose among various email service providers by looking at the segmentation service they offer.

New email subscribers

Your new subscribers don't know your services. Hence, you can't approach them with the same system you follow for the old ones. It is important to offer them valuable messages to prepare as a customer. You should segment new subscribers and generate a separate email approach to convert them into customers. You can start by sending a welcome message along with a short plan of what you have for the subscribers in coming weeks. Then, keep sending valuable emails with resources that match your subscribers' need.

Interest segmentation

With a reliable segmentation tool, you can have a huge dataset associated with subscribers' interests. These interests prove a great segmentation approach to enhance conversions. For instance, you can offer weight loss products to subscribers who have a great interest in the fitness-related content. Or, provide email fitness courses to the segmented group.

Service preference segmentation

Your email list includes your customers. And they need to feel happy with your email services. Understanding customer preference is highly important to keep them happy. In email lists, customer preference includes types of content, the frequency of emails, discounts, and other updates. You should describe different segments of subscribers according to their preferences. This will save you from losing subscribers.

Rate of engagement

In your huge email list, you will experience that there are people who consistently engage with your messages. However, you will also find a group that rarely opens. Your software will tell you about the open rate. You need to segment subscribers into three categories.

- High open rate
- Medium open rate
- Low open rate

This segmentation allows you to know which users can be pushed towards high rate. You can present rewards for medium open rate subscribers to gain more engagement from them.

Similarly, you can generate an email plan for low open rate subscribers too. These are inactive ones, so you need to remind them about your services. For instance, if you offer exclusive courses, send reminders to people who are not following weekly course resources.

Lead segmentation

When signing up, every subscriber opts for different lead magnets. These lead magnets are the type of services or products they desire. Lead segmentation will simplify the process of messaging and promotion. You can create a group of subscribers who desire PDFs, webinars, downloadable articles or other types of content. Thanks to automated tools, this process won't take too long.

Opened, but not converted

You will always find a difference between open rate and conversion rate. Many people open emails, go through the message but don't follow the conversion process. You can create a segment of such subscribers and let them know about the missed-out opportunity.

You can start by sending a detailed message with valuable information related to your product. It is possible that those people weren't satisfied with the previously delivered information.

So, you can introduce your product in a detailed manner to get more conversions.

Purchase history

This segmentation is highly essential for marketers who promote multiple products via emails. Some people buy same products again and again, while others choose different products every time you promote. Knowing purchase history will help you reduce your efforts and improve results. You will know which list member would be ready to buy your product. This will save from making unsuccessful efforts.

Buyer's cycle and frequency

People who regularly buy from you should get rewarded. This brings loyalty to your email marketing. You can send an invitation to your frequent buyers to join new campaigns and get benefited.

Similarly, you will find higher conversions in different time intervals. This is called buyer's cycle, which depends on your product and buyer's need.

When marketing through emails, you can design a message delivery system that matches your buyers' cycle. This increases the chances of conversions, as you present your offer right when your subscriber needs them.

Desktop and mobile customers

Most email marketers don't realize, but this market segmentation is extremely important. An email looks completely different on mobile due to smaller screen size. It is prudent to fathom the number of subscribers, who read your emails through smartphones. Then, you can design email copies for high readability and target accordingly.

Try to find responsive email templates that look appealing on a desktop as well as mobile.

Subscribers who help in promotion

As you grow your email list, your subscribers start referring your email services to others. You should take special care of these subscribers. You can send a "thank you" message along with rewards regarding free content. This will encourage them to keep on sharing your content with others and grow your email list.

The profession of your subscribers

His or her career majorly defines a person's interests. A content writer looks for content resources. Similarly, a physical trainer looks out for offers on fitness products and equipment. Understanding and segmenting your subscribers concerning their profession is a great move. You can find different selling points and address problems with your products and services as a solution.

Gender segmentation

If your marketing strategy is directed separately towards men and women, you need this segmentation. Define gender groups in your email lists and design focused messages for both men and women.

Segmenting for education level

For some email marketers, education level of a subscriber matters. For example, if you desire to sell multiple courses on marketing, you should know the education level to send right emails. You can't expect a beginner marketer to respond to an advanced level of course. Hence, it is essential to divide your email lists into different education level categories and send course-related emails that are most suitable for a group of subscribers.

Segmenting according to income

Your subscribers choose products and services according to their buying capacity. You can understand that buying capacity by knowing their average income. Define different ranges of income category and put email subscribers in those categories. You can even go up to 5 figures or 6 figures incomes, depending on what kind of product you want to market.

Every member of your email list behaves differently. It is your job to understand those behaviors and modify your marketing approach time to time. Segmentation through automated and manual processes can unlock the maximum potential of passive income through email marketing.

Qualifying your Subscribers

Before you begin your message distribution, make sure that your subscribers want your services. Qualifying emails means that you inform your contacts about your message distribution plan. This way, you don't send emails against any subscriber's will and don't seem like spam.

There are many different ways to qualify your email list and ensure that they are your opt-in subscribers.

How to qualify your subscribers?

Is your email list legitimate and qualified for marketing?! That depends on the methods you leverage to obtain your list.

Site visitors who permitted you to send emails

These are your opt-in email list members. They know about email services at the time of accepting your services. They sign up for your emails by giving their permission on your site. However, it is also vital that you clearly state what you want to do with their emails; if you wish to send newsletters or desire to include them in marketing campaigns.

It is advisable to confirm such subscribers by sending a confirmation message. This will ensure that they qualify as your opt-in subscribers.

Subscribers who filled a form on your landing page

If a subscriber fills all the information asked on a landing page, he or she qualify as a subscriber. But the form should explicitly say that you are asking for emails for campaigns. There are automated tools you can use to design your landing page and get subscribers.

Such subscribers usually know and understand the services. Hence, you can include them in your opt-in collection. Some marketers like to get a second confirmation by sending a clear email including the details of campaigns you plan to run.

Business cards you obtain from office visitors and trade events

Generally, these people don't know if they are on your email list. So, you can't count them in your opt-in collection. The right way to incorporate these contacts is to send a series of emails in a non-pushy way. Describe what you are trying to accomplish and benefits that you can offer. Then, it is up to those people whether they accept or not. Keep adding people who confirm the acceptance of emails.

Members of your social media community or forum

Sure, these individuals are more likely to accept your emails. But you can't include them in qualifying list without getting their permission. A subscriber should know and have the necessary understanding of your campaigns. You need to contact these members and ask whether they would like to become a subscriber or not. Try posting a link to your landing page and direct these people to the subscription form. This will make them a legitimate contact.

Customers who have bought a product or a service from you

Depends on how they bought your product or services. If the medium was email, then, you can count them in your opt-in collection. However, products sales via social media, site, or any other method requires follow-up. You can suggest customers that they can get more services or products time to time via your email services. Get their permission to use their email IDs. Then, you can keep sending them new pitches.

Emails available in your address book

During list building, you can include all contacts available in your address book. However, they don't necessarily qualify as your subscribers. Send your marketing proposal in an email and ask for confirmation. If they accept, then, you can include them in your opt-in collection.

The list you have rented or purchased

If you have rented a list from another email marketer, they don't qualify as your subscribers. You can use them to increase your email list. But it is critically important that you treat them as potential subscribers. Send valuable messages to engage them. Then, offer content useful to pitch your products finally.

Clean your email list

After qualifying your subscribers, you can follow a few necessary steps to clean your list. Cleaning your email list includes different methods to enhance the quality of emails.

- **Get rid of misspellings**

First of all, you need to resolve misspelling and typos. Collect email IDs that bounce back whenever you send messages. Review these IDs to find typos and resolve them.

Also, check which emails don exist any longer. These IDs have nothing to offer, so it is wise to remove and try to look for new emails from the same subscribers or others.

- **Find inactive emails**

Sometimes, correct email IDs stay inactive too. You can find these emails when checking data associated with an open rate on a monthly basis. This data will allow you to gain insights about inactive subscribers. People who don't even open your email for months require special attention. You can create a special campaign for such subscribers and ask for their needs and interests.

For inactive emails, you need to create an appealing subject title to grab their attention. An offer or a discount can help in this process too. It will turn many inactive emails into active ones again.

Building a Relationship with Your List

An essential aspect of email advertising is building a healthy relationship. Sure, you get a better response via emails, but that relies on how you connect with your subscribers. This is a fast-paced marketing world. Things move fast, and people don't have too much time to offer to any random marketer.

As an email marketer, it is wise to use high-quality technologies. However, you should keep the basics of marketing in mind. Finding a potential customer is one thing but keeping that customer on a regular basis requires consistent efforts.

Whenever thinking about relationship building, you should concentrate on three major factors- **awareness, likability, and trust.** Achieve these factors, and you become a successful email marketer.

How can you achieve all three factors?!

Start with message personalization. A personalized email connects you directly with your email list. Tapping into email marketing is a success only when you create strong relationships with your actions. Hence, utilize all segmented data associated with education, location, gender, and others to make your messages more personalized. Keep an eye on consistent and rare open rates. Define subscriber-oriented message strategy. That is how you win every subscriber and develop a strong, long-term relationship.

Become a consistent and disciplined email marketer

Your relationship with email subscribers is similar to your general social relations. You regular spend time and interact with your friends and family via calls, online and offline communication. The same approach helps in email marketing as well.

Your email list will grow concerning loyalty if you stay connected. Your emails should reach your list consistently. You can define a disciplined system on a weekly and monthly basis. However, the key is to stay stick to that system. Only then, your messages will receive a high level of attention.

Share a variety of content that matches subscribers' interests

Your subscribers should trust you as an expert resource. So, you need to become a one-stop solution for relevant content. Combining text messages with appealing media, videos, audio and other sorts of content would be a great marketing move. However, they all work when you only keep the interest of your email list in mind. Choose different ways to present relevant information. Eventually, your email subscribers will start looking at you as a reliable solution. Then, you can think about product promotions and selling items.

Make them click every email with a dazzling headline

You can connect to your email list only when they open your emails. And that majorly depends on how you write your message headlines. A dazzling subject looks different than other messages in an inbox. It attracts people with a clear, crisp and beneficial message.

Here are a few examples to create click-worthy subject lines:

- Use brackets to highlight special features of your emails. You can write [FREE EBOOK] or [VIDEO CONTENT] to grab the attention of your subscribers.
- Use emojis and numbers to write more attention-grabbing subject line.
- Present your subject as a question to generate curiosity.

You should consistently experiment with new types of subject lines. This will evolve your techniques to gain higher open rates.

Optimize emails for user convenience

Email users differ regarding priorities, platforms, and devices. Your job is to understand those differences and optimize your messages accordingly. As mentioned earlier, optimization for mobile devices is the most important part. Maximum conversions come through mobile users. When you think about your users, they start responding more. Make emails convenient to read on mobile, and you will see better loyalty rate from your email list.

Many users don't allow images on their emails. You can include ALT text in your images for such users. So, even without the images, your user will see the name of that image along with a clickable link. Similarly, you should define goals of your subscribers and incorporate them in your email copy. This will reduce confusion at the time of message production. Plus, you will deliver exactly what your users want to read.

Don't just try to sell

If you try to sell things, soon your subscribers will stop opening your emails. The whole idea of finding and segmenting email lists is to provide value to your target market. You should always divide your email campaign into two parts. One part should cover content and resources to help audiences. And the rest part should focus on selling products or services. Smart marketers follow 80 and 20 percent parameter when dividing marketing approach. Larger part goes to providing valuable resources.

With this approach, you connect with your customers on a personal level. A relationship generates concerning mutual interests. And that is how you sell more in future. Regular valuable resources make your promotional pitches more effective. Subscribers pay attention to your emails and respond by buying those products.

Combine multiple affiliate deals in emails

If segmented smartly, you can promote numerous affiliate deals in one email. This improves the chances of boosting your profitability. But it also presents you as a reliable source in front of your email list.

Choosing the right affiliate deals is the key here. You should combine non-competitive products, so that, they don't cancel each other's solutions. Focus on your content first. Point out different problems through your email content. Then, present complementing affiliate offers to resolve those problems. Just remember that those deals are desirable and match your target market's need.

Write interactive call-to-actions

The most significant mistake that marketers make is following a formal call-to-action format. Multiple lines, in the end, asking for different actions make you look too profit-oriented. The end of your email should have an interactive touch. The call-to-action should include everything you need from your subscribers. However, the writing format needs to have a light communicative model.

For example, you can write, "A great course is waiting for you on the other side, click now to start growing."

This line includes all standard features of a call-to-action:

- A link to click to
- A crisp sentence
- A pathway towards making that purchase

But that is not all! You can see how light and interactive this call-to-action is. And this property further enhances the chances of getting conversions on a regular basis.

Use impressive rewards

Almost every email marketer tries to impress people with discounts and rewards. Subscribers receive hundreds of emails with such offers. But you can create loyalty in your email list with unique and relevant rewards.

Always think about your segments' interest before deciding rewards. For instance, long-term subscribers would love to get rewarded for their loyalty. Similarly, you can present special rewards for word-of-mouth for those who promote your campaigns.

Always publish on-time

Email marketing allows you to define your approach to message delivery. However, that doesn't mean you should follow a random publishing system. Your subscribers should be able to put their faith in you. There are some marketers who lose their email list due to not providing resources on time.

You can design your email distribution frequency according to segments. But then, make sure that those messages reach subscribers in the same manner. Your audience will be aware of the fact that 'when to expect your emails,' which will positively enhance your open rates and conversions.

Thanks to email distribution tools, you should feel any difficulty in following a disciplined format of publishing.

Avoid sending too many emails every day

If you want them to respect your marketing approach, you have to respect their time as well. Ultimately, your email list is a list of people who have their own lives and daily responsibilities. It is critically important for you to recognize that and respect that. Don't send too many messages on a daily or weekly basis. A constant reminder of the same thing can cost you many subscribers.

Design a balanced system that fulfills your marketing plan and keeps your subscribers happy too. A maximum of 2 times a week is a good system to follow; and increasing your frequency for those who engage in your emails.

Always send easily consumable information-packed emails

This is another aspect of not wasting your subscribers' time. You need them to take your messages seriously. And that won't happen if you keep on sending unnecessary messages with too long bulky content.

Information is important. However, a smart presentation of that information is much more valuable for email marketers. A person gives a few seconds to an email, so you need to point out very effective components to make an email work. Write an email copy that enables a quick scan through naked eyes.

Short sentences, bullet points, crisp concepts, these are the elements you need in your format. It gives a whole lot of information in a very short time period. Hence, your subscribers feel more comfortable opening and reading your emails whenever you send.

Choose your autoresponders wisely

An autoresponder saves from spending all your time replying to new subscribers. It immediately sends an automated message to new people who get added to your list. Make sure that you choose a reliable and swift autoresponder. Also, ensure that your automated responses are warm and engaging. This will create a strong first impression and set a foundation for future communication.

Choose a casual tone of conversation when designing such emails. The language should be welcoming and appropriate as well. The idea presents you as a friendly and credible source. Then, you can include this email in your response system.

Monitor results of your efforts and modify

A strong relationship is not a one-day job. You have to regularly improve your abilities and interact with subscribers to gain their trust. Consistent measurements can help you do that. Keep an eye on multiple email metrics to understand how engaging your efforts are. Give additional focus to open rate, CTR, and conversions. This will offer insights associated with trending interests, most successful segments, and high-rewarding campaigns.

Evaluating data regularly will set a foundation for relationship management. Then, you can use these insights to generate different scenarios. Try a different approach to see how your email list responds and improve accordingly. Make sure you include sales and leads when estimating your overall ROI.

Focusing on every step mentioned above, you can chase your profit goals without losing the trust of your email subscribers. Your emails will become appealing for all kinds of segments no matter which device they use. Consistency and balanced frequency will obtain you an undying curiosity. Hence, you will develop a regular communication channel with your messages.

Make Money with Your List

Here are some unique ways to make money with your well-segmented email list.

Indulge in affiliate marketing

There are many brands and businesses out there offering email marketing affiliate commissions. Such affiliate programs allow you to market products and gain commissions in return. This approach is highly popular for passive income generation. Any person can create his or her email list and incorporate affiliate programs to make money. You can easily earn up to 25 percent of overall product price commission. A simple inclusion of affiliate link in your emails is enough to let the magic happen.

Marketing a product is easy. The real critical part is selecting products for affiliate marketing. You need to find products that are relevant to the interest of your subscribers. Plus, those products should have affiliate programs too. Brand power also matters if you come across two options in the same product category.

With in-depth analysis, you can choose a perfect product for affiliate marketing.

Offer advertising space in your emails

When you successfully build a large email list, it is possible to run ads on your newsletters. You need to have more than 2500 subscribers to make it happen. Email marketers charge great amounts to provide a small space for ads.

An email with advertisement includes several lines associated with that product along with a link.

Sell your products

You can sell your products such as courses, books or others, depending on your expertise. The best approach to selling would be presenting your product to your subscribers. Promoting your product is not about commissions. You need to try and create a continuous cycle of sales, which is a little harder in this case.

First, start by clearly defining what your users want. List segmentation will help you do that. Plus, you can directly ask your users about what sort of products they would like to have. Direct their attention towards a problem and present suitable solutions. This will allow you to understand which solution is most trending among your users.

Use answers to design your products in a mold that impresses the most. It will surely take plenty of time, and in some cases, money as well. You can take over the whole product generation process to save from high expense.

When you are designing your product, remember that quality is the most important factor. It doesn't matter if it takes longer, you should never compromise on quality. Pay attention to details and regularly match product's growth with market needs.

Finally, you can start using relationship building techniques to promote your product.

Rent email lists to others

Not very common, but many email marketers use this technique to earn. You can get quick money by renting your email lists to other marketers and businesses. However, it all comes down to your preferences. Most marketers spend months in generating and optimizing their lists. So, you may desire to keep that list for your marketing goals only. But if you do choose to rent, make sure that you rent to a legitimate and credible source.

Follow a proven promotion path

As you keep trying multiple ways to lock-in customers, some campaigns will thrive, while others won't. A campaign that has given you positive results in the past is worthy of a replay. You can gain the same excitement and sales with repeated email marketing campaigns.

The process is simple. You need to keep an eye on campaign performances. Notice sales and ROI of all your campaigns and align them in descending order of success. This list will help you choose which campaigns are perfect to try again. A product with discounts or special promotions can make people pay instantly. So, you shouldn't stop yourself from trying proven methods again and again.

However, make sure you don't make your campaigns look too repetitive. Balance in such a manner that your subscribers look at your campaigns as a fresh deal. And that will bring you positive results every time you try an old campaign.

Match sales approach with list segmentation

You already know how campaign segmentation can skyrocket your marketing ability. But it depends on how well you choose products for every segment. Segments associated with open rates, demographics, past purchases, product interests and others are important here. You can use these segments to define pitch force and type of products. Email subscribers will see products that they look for. Hence, the conversion rate will increase for sure.

Use an up-selling approach

Up-selling is when you top one product with other at the time of selling. An email subscriber accepts a product, and then, you offer another complementary product to enhance the price. It is essential for the second product to work as an upgrade for the first one you have sold.

Think about how a restaurant service asks you for beverages when you order a burger. Up-selling is the same. In your case, second products can be upgrades or even additional services. The product can change with industries, but they all come under the concept of upselling.

Use down-selling approach

Sometimes customers accept then back out without purchasing the product. This is common, as people feel uncertain about their investment in products. Your job is to make that deal more valuable for them to get them to buy that product.

If possible, try to lower the price, so that, they can buy. This process is called down-selling. Make sure you make that reduced price look like a one-time offer. This will create a sense of excitement and increase the chances of conversion.

Re-target "opened, but not converted" segment

Most of your marketing efforts can face low conversion rates due to this problem. However, you are not the only one. Every online business has to fight this problem. You can copy their approach in your email marketing.

People who open your emails, but don't buy have some issues with your product. Your job is to resolve those issues by offering a series of valuable emails. This is called re-targeting. You provide detailed product information, present discount opportunities and remind them about the benefits of the products.

Re-targeting can increase your conversion rate up to 50%. Campaign's CTR will increase, and you will see a boost in sales as well.

Sell premium content through a paid subscription

Do you have expert knowledge in your field?! Then, your email list can become your market. You can develop a vast collection of premium content and sell them through paid subscription.

To attract subscribers, you need to provide them free content. Offer valuable content pieces and include a call-to-action informing them about the paid subscription. You can provide one-time content, monthly or even annual membership to your subscribers. Dividing into different formats will allow most of your subscribers to choose paid content, according to their preference.

Even with a small payment, you create a consistent cycle of revenue. Your content is there forever, so all you need is new subscribers to increase your sales.

Start one-on-one consulting or coaching

Email marketing is a great way to build a strong consumer base. Your consistent help and suggestions offer benefits to subscribers. As a result, customers start relying on your content.

Then, you can pitch your coaching or consulting services. These services usually include having one-on-one interaction with subscribers. That is why you get paid way more than content selling.

Offering to coach via social media or any other platform doesn't give you that much exposure. However, emails create a sense of personalization and security. Hence, people are more likely to connect to get your help on a regular basis. You can divide your sessions according to your free time and make money.

It is necessary to remember the importance of quality here. Your suggestions need to be effective and practical. A customer should be able to use your coaching to generate desired results. Only then, you can expect continued growth in your market base.

Choose your coaching content very carefully. You should have experience and effective knowledge in that field.

Rules that will help you make money consistently through your email list

Till now, you have received multiple methods of making money with email marketing. But the success of those methods is possible when you follow certain rules:

Rule#1 Pick your niche and stick to it

No business can achieve success without picking a niche. Your customers should know what you are selling. Hence, you need a niche. Trying to pull in customers in different categories can limit your ability to grow. It usually confuses your subscribers, and you don't get to define your marketing approach.

There is nothing wrong with selling multiple products. However, all those products should belong to the same line. For example, you can sell cooking recipes, cooking appliances, and cooking coaching classes altogether. The idea is to target a certain market and segment according to their needs.

Selecting a niche makes things easier for you and your subscribers as well. People feel clear when they decide to sign up as a subscriber. Otherwise, your services seem all over the place, making people think twice before opting for products or services.

It is critical that you choose your niche according to your email list. Hence, the process of niche selection should start way before list building and segmentation. This way, you become familiar with what your market needs. Then, align those needs with your expertise and apply in list building.

Rule#2 Make your pitch valuable

You have read the term "valuable pitch" for many times in this book. But what does that mean?! A valuable pitch matches the needs of subscribers in a unique manner. Marketers who make money don't write their sales pitch focused towards product features.

To write a valuable pitch, you need to understand the **difference between product features and product benefits.** A potential customer responds to product benefits, rather than product features.

You can't focus too much on your product. Of course, your purpose should be to sell your product. But that is possible when you present that product as a solution. People think about their problems whenever they see a product. If they don't, conversion doesn't happen. Your job is to remind about a problem and present solution through your pitch. That is how you can make your pitch valuable.

Rule#3 Define a path towards sales

A content piece should not just be a content piece. Every email, message, and resource you provide to your subscribers should align in one path. And that path should lead to sales.

With blogs, articles and other content, you can make your services valuable. So valuable that your subscribers start waiting for your emails. Then, you start pitching and upgrading your emails from resources to sales pitch. Inviting subscribers to buy a paid product, then, works effectively. Hence, you can sell your eBooks, courses, items and other products.

All in all, every content piece should have a purpose. So, it is not just about enhancing the value. It is about enhancing the value in a certain direction. That is how you make money through email lists.

Many marketers become pushy and send spam emails. This happens when a marketer doesn't define a clear email sequence. Overdoing pushes customers away. You have to stay away to increase email sales. Treat your subscribers as your friends and provide with genuine offers via your email sequence. Present your offers in multiple manners without overdoing it.

How to convert every subscriber into a customer?

You complete 50% of the job by finding the right emails to add to your list. However, the rest of the job relies on your ability to turn those subscribers into customers. You already know various ways to make money via email marketing. But it is also necessary to ensure that every subscriber becomes a customer. This approach will lead you to maximize revenue.

Here is what you should do:

1. A well-defined sequence

A successful email sequence includes following steps:

- An email that congratulates subscribers for the subscription
- An email that tells stories of happy subscribers.
- An email that offers products along with links and CTA.
- An email that presents deadlines to purchase products for maximum benefit.

- An email that reminds final moments to grab an amazing deal.

You can incorporate an autoresponder to create such a system. There are many paid tools available that you can find and leverage.

2. Make buying easiest through your service

There are all kinds of products available on the internet. So, the question arises- why a person should buy products from you?!

You can present speed of purchase as your answer. Of course, the quality of products or a service matters. But it is necessary that your subscribers obtain a smooth buying experience. Hence, you need to be precise when creating your sales emails. Also, optimize those emails for all kinds of devices.

Try to include as few redirections as possible and provide a common payment model. This will help customers to buy from you on a regular basis.

Conclusion

As an email marketer, your email list becomes the best asset you can have for passive income. It is your list that helps in building relationships, finding regular customers and making money on a regular basis.

No matter how many social media platforms are launched, emails are going to be valuable forever. However, just building a list of emails can't provide value. You need to find the right types of subscribers to make marketing profitable. So, your major goal is to remove irrelevance from your list. Then, the rest of your subscribers will become your long-term customers.

Keep on reminding why you are building your email list. You ultimately want to make money with your emails. So, focus on subscribers who you can turn into customers in the future. Make your emails compatible with latest mobile technologies to become a vital source.

When generating freebies for subscribers, don't give everyone the same option. Some might like a checklist, while others will look for an eBook.

To engage subscribers, understand their buying intention. Then, create a small piece of a product as a free model. This will help your subscribers and make them more interested in your paid products. Remember, freebies are not to get more subscribers. They are provided to lead towards product pitch.

Here is a summary of every step you need to accomplish as an email marketer:

- Spend hours researching your target market to know their likes and dislikes.
- Use the obtained conclusions to define product needs that you can fulfill.
- Leverage multiple techniques and tools to build your email list. Focus on getting a targeted list that will prove useful for your product or service idea.
- Learn about segmentation and gather resources to start segmenting. Choose segments that are more valuable to campaigns you have in mind. You can even define multiple segments for different products you have.
- Improve your email list by qualifying every subscriber. Make sure every subscriber knows and wants your services.

- Design a value providing email sequence for different segments. Constantly stay in touch with your subscribers via emails that offer quality resources.
- Choose different methods to sell your products or affiliate products. Use long-term techniques to keep on making money from your emails.
- Follow an automated sequence that works on its own, so that, you can keep on passively getting revenue.
- Learn all about email marketing metrics and gather reports through analytics. Keep monitoring open rates, conversions, and other metrics regularly. Modify your email campaigns according to the available data and maximize ROI.

All in all, email list building and making money depends on how valuable you are in the eyes of your subscribers. The more you offer, the better you become.

So, start building your email list and invest time in gathering effective marketing tools. The initial phase requires dedication and patience. But then, you will create an automated cycle of passive income.

Share your knowledge with others and help them earn as well!

References

http://www.marketing-schools.org/types-of-marketing/email-marketing.html

https://www.roimachines.com/10-email-marketing-benefits/

https://www.pure360.com/10-benefits-of-email-marketing/

https://www.entrepreneur.com/article/285949

https://www.verticalresponse.com/blog/40-brilliant-but-easy-ways-to-build-your-email-list/

https://socialtriggers.com/list-building/

https://www.dailyblogtips.com/how-to-build-an-email-list-that-makes-money/

https://optinmonster.com/50-smart-ways-to-segment-your-email-list/

https://blogs.constantcontact.com/profitable-email-marketing/

http://traceylawton.com/blog/build-your-relationship-with-your-subscribers-with-these-7-simple-ways/

https://blog.leadquizzes.com/12-clever-ways-to-make-more-money-from-your-email-list

https://wppopupmaker.com/email-marketing/make-money-with-your-email-list/

https://www.melyssagriffin.com/make-money-email-list/

https://help.activecampaign.com/hc/en-us/articles/206907550-What-is-an-opt-in-list

www.ingramcontent.com/pod-product-compliance
Ingram Content Group UK Ltd.
Pitfield, Milton Keynes, MK11 3LW, UK
UKHW021259180426
11947UKWH00015B/927